Donnelly, Karen J.

Ice ages of the past and the future.

DATE			

Ice Ages of the Past and the Future

Karen Donnelly

The Rosen Publishing Group's
PowerKids Press™
New York

To my family: Cathy, Colleen, and David

Published in 2003 by The Rosen Publishing Group, Inc.
29 East 21st Street, New York, NY 10010

First Edition

Editor: Gillian C. Brown
Book Design: Michael J. Caroleo

Photo Credits: Cover and title page © CORBIS; page borders, p. 8 © Weatherstock/Warren Faidley; p. 4 © Emily Muschinske and Shae Chatlin; p. 7 © Rear Admiral Harley D. Nygren, NOAA Corps (ret.); p. 11 illustration by Michael J. Caroleo; p. 12 © Gunter Marx/CORBIS; p. 15 © Richard Cummins/CORBIS; p. 16 © Michael Van Woert, NOAA NESDIS, ORA; p. 19 © Michael Pole/CORBIS; p. 20 © Morton Beebe, S.F./CORBIS.

Donnelly, Karen.
Ice ages of the past and the future / Karen Donnelly.— 1st ed.
p. cm. — (Earth's changing weather and climate)
Includes bibliographical references and index.
ISBN 0-8239-6219-9 (library binding)
1. Glacial epoch—Juvenile literature. 2. Global warming—Juvenile literature. [1. Glacial epoch. 2. Global warming.] I. Title.
QE697 .D59 2003
551.7'92—dc21

2001007874

Manufactured in the United States of America

Contents

Amazing Glaciers

About 18,000 years ago, during the Pleistocene Ice Age, North America looked very different from how it looks today. Over thousands of years, **glaciers** slowly expanded south. Ice sheets covered the ground from Canada as far south as Pennsylvania, Ohio, Indiana, and Illinois. Sea levels were about 430 feet (131 m) lower, because so much of Earth's water was trapped in the glaciers.

Temperatures warmed, causing the glaciers to melt. Glaciers made amazing changes to the land. For example, glaciers scraped the Great Lakes into the land that surrounds Michigan. Yosemite Valley in California's Yosemite National Park was carved by a glacier. In North Dakota, glaciers created 80-foot-high (24-m-high) sand dunes.

At Glacier Bay National Park and Preserve in Alaska, the different layers of compacted ice and snow that make up a glacier are easy to see.

What Is an Ice Age?

An ice age is a period of time when the temperature of Earth is very cold. During an ice age, glaciers expand to cover more and more land. A glacier is made up of snow and ice that does not melt. When snow piles up, the weight of the snow on the top presses down on the snow on the bottom. The snow on the bottom is pressed tightly together and becomes ice. Glaciers get bigger as even more snow piles up. They expand as the top layers slide downhill along the bottom layer of ice.

Scientists believe that an ice age occurs when changes in the **climate** act together to cause temperatures to cool. Many ice ages have occurred throughout Earth's history. Earth's orbit, along with several other elements, influences ice ages.

Three-fourths of all freshwater in the world is frozen inside glacier ice!

The Earth's Orbit

Earth orbits the Sun once each year. Earth's orbit affects the amount of energy, or heat, that reaches Earth from the Sun. As Earth orbits the Sun, it **rotates** on a tilted **axis**, like a spinning top. When the Northern **Hemisphere** tilts toward the Sun, it has warm, summer temperatures. The Southern Hemisphere is tilted away and has cold, winter temperatures. Every six months, these conditions switch.

The angle of Earth's tilt changes from about 21 degrees to about 25 degrees over a period of 41,000 years. A smaller angle brings heavier snowstorms to the polar regions. Glaciers and ice sheets grow because of the increase in snow. More snow reflects more of the Sun's heat energy away from Earth and causes additional cooling.

Earth is closest to the Sun in January, but it is winter in the Northern Hemisphere because of Earth's tilt.

The Milankovitch Theory

In 1938, a Serbian scientist named Milutin Milankovitch described how the orbit of Earth around the Sun affects climate. He saw that over thousands of years, the orbit changes in two ways. These changes affect the way the Sun warms Earth.

Milankovitch explained that over a period of nearly 100,000 years, the **elliptical** path of Earth's orbit changes slightly. This changes the distance between Earth and the Sun. When Earth is farther away from the Sun, the climate is cooler. He also described how Earth's angle changes over a 41,000-year period, as we learned in the last last chapter. These cycles determine when Earth's orbit will be farthest away from the Sun and the climate will be cooler. Some scientists believe these changes cause ice ages.

This picture shows the Earth rotating on the elliptical path around the Sun.

Why Are Orbit Cycles Important?

Considered alone, these individual cycles do not have a great effect on climate change. The combined effect of the angle of Earth's axis, the elliptical path of Earth around the Sun, and the direction of Earth's tilt are important in determining when an ice age is likely to occur. Sometimes the warming effect of one cycle cancels out the cooling effect of another. However, when the cooling effects of all three cycles occur at the same time, an ice age would be predicted. Using the mathematical **calculations** developed by Milankovitch, some scientists predict that an ice age is not expected in the next 100,000 years. Other factors, such as **global warming**, also make it unlikely that an ice age will occur in the near future.

Global warming is due in part to human activities, such as driving cars and cutting down trees.

Temperatures on the Rise

Some scientists believe Earth is getting warmer because of an increase in **greenhouse gases**, such as **carbon dioxide**, in the **atmosphere**. Global temperatures have increased about 1°F (.5°C) over the last 100 years. This is 10 times faster than in the past!

The gases in Earth's atmosphere work like the glass walls of a greenhouse. Light and heat from the Sun pass through the atmosphere to Earth. The land, water, plants, and buildings soak up energy from the sunlight. Some energy is sent back toward the Sun, but much of it is trapped by the atmosphere. This causes the Earth's temperature to rise. Higher levels of carbon dioxide increase the greenhouse effect. Scientists are concerned that people are largely to blame for global warming.

Carbon dioxide is a greenhouse gas that is released into the air when we burn fossil fuels, such as gas, oil, or coal.

Humans and Global Warming

During the **Industrial Revolution**, humans began burning a lot of **fossil fuels** to make electricity and to run cars. Since then the level of carbon dioxide in the atmosphere has increased between 24 and 30 percent. Scientists think this may cause global warming.

Warmer temperatures melt polar ice sheets, sending more water into the oceans. Sea levels, the places where oceans touch land, may rise. Some scientists are worried that if water temperatures around the South Pole warm a few degrees, the West Antarctic Ice Sheet could break apart and slide into the sea! If that happens, sea levels could slowly rise as much as 20 feet (6 m). To help predict future climates, scientists gather information about Earth's climate history.

If all the existing glacier ice melted into the ocean, the world's sea levels would rise 300 feet (91 m)!

Climate History

How do scientists find out Earth's climate history? They look for clues in plants, on lake bottoms, and on the ocean floor.

Annual **growth rings** in trees tell the climate's story. Wide rings indicate a year with a lot of moisture. During a dry year, the ring is thin. **Sediment** found in the bottom of a lake includes pollen and seeds from plants that grew on the banks. Studying these clues shows scientists whether forests or grasses covered the area. Ocean sediment reveals **fossils** of tiny animals that lived many millions of years ago. The species of plants and animals able to live in an area gives scientists information about temperature and moisture. All of this information helps scientists to know what the temperatures were and how much **precipitation** fell in the past.

Scientists who study Earth's changing climate are called paleoclimatologists. They get clues from many sources, such as trees.

Earth's Frozen Past

Perhaps the most important tool that scientists use to study global temperature and when ice ages occur is a long tube of ice called an ice core. Scientists drill deep into the glaciers of Greenland and Antarctica. They pull out an ice core that is about 4 inches (10 cm) across and as much as 2 miles (3 km) long. Almost like the growth rings in trees, the ice core is made up of tiny layers. There is one layer of snow for each year, and each layer provides information about climate. Fossilized air is tiny air bubbles trapped in the ice core. These help scientists learn what gases were in the atmosphere when each layer of snow fell. Scientists have found higher levels of carbon dioxide since the beginning of the Industrial Revolution.

The simplest ice core drill is called a hand auger and can retrieve an ice core sample that is between 98 and 131 feet (30–40 m) long.

The Land Tells a Story

After a glacier melts, changes it made to the landscape tell how the ice came and went. Valleys are cut through mountains. Huge boulders, too large to have been brought by running water, are carried many miles (km) and set down by glaciers. By studying the kind of rock in the boulder, scientists can tell where its journey began.

Scientists believe that over Earth's lifetime there were periods when there were no polar ice caps at all. Today ice caps in the Arctic and Antarctica seem normal. We are in an interglacial period, a time when glaciers shrink as the climate warms. Scientists can use Earth's climate history and orbit cycles to predict that a new age of expanding ice is unlikely in the next 100,000 years. However, no one can really be sure.

Glossary

atmosphere (AT-muh-sfeer) The layers of air surrounding Earth.

axis (AK-sis) A real or imaginary straight line through the center of an object, around which an object turns.

calculations (kal-kyoo-LAY-shunz) To find out by using addition, subtraction, multiplication, or division.

carbon dioxide (KAR-bin dy-OK-syd) A gas that plants take in from the air and use to make food.

climate (KLY-mit) Average weather conditions over a long period of time.

elliptical (ih-LIHP-tih-kuhl) Somewhat egg-shaped.

fossil fuels (FAH-sul FYOOLZ) Fuels such as coal, natural gas, or gasoline that were made from plants that died millions of years ago.

fossils (FAH-sulz) The hardened remains of animals or plants that lived long ago.

glaciers (GLAY-shurz) Large masses of ice that move down a mountain or along a valley.

global warming (GLOH-bul WAR-ming) A gradual increase in the temperature on Earth.

greenhouse gases (GREEN-hows GAS-ez) Different gases in the atmosphere that keep heat from escaping from Earth.

growth rings (GROHTH RINGZ) Layers of wood on a tree that mark growth. One layer stands for one year's growth.

hemisphere (HEH-muh-sfeer) Half of Earth's surface.

Industrial Revolution (in-DUS-tree-uhl reh-vuh-LOO-shun) Changes in the way people worked in factories and on farms, beginning in the mid-1700s.

precipitation (prih-sih-pih-TAY-shun) Any form of water that falls from the sky.

rotates (ROH-tayts) Turns around on an axis.

sediment (SEH-dih-ment) Gravel, sand, silt, or mud carried by wind or water that settles to the bottom of lakes, rivers, and oceans.

Index

Web Sites

Due to the changing nature of Internet links, PowerKids Press has developed an online list of Web sites related to the subject of this book. This site is updated regularly. Please use this link to access the list:

www.powerkidslinks.com/ecwc/iceage/